DATE DUE

[redacted]

[redacted]

[redacted]

[redacted]

[redacted]

Acid Rain

Locating the Author's Main Idea

Curriculum Consultant: JoAnne Buggey, Ph.D.
College of Education, University of Minnesota

By Robert Anderson

OPPOSING JUNIORS VIEWPOINTS®

Greenhaven Press, Inc.
Post Office Box 289009
San Diego, CA 92198-9009

Titles in the opposing viewpoints juniors series:

Acid Rain
Addiction
Advertising
AIDS
Alcohol
Animal Rights
Causes of Crime
Child Abuse
Christopher Columbus
Death Penalty
Drugs and Sports
Elections
Endangered Species
The Environment
Forests
Free Speech
Garbage
Gun Control
The Homeless
Hunger
Immigration
Nuclear Power
The Palestinian Conflict
Patriotism
Police Brutality
Pollution
Population
Poverty
Prisons
Smoking
Television
Toxic Wastes
The U.S. Constitution
The War on Drugs
Working Mothers
Zoos

Cover photo: © Greenpeace/van der Naald

Library of Congress Cataloging-in-Publication Data

Anderson, Bob, 1950-
Acid rain : locating the author's main idea / by Robert Anderson ; curriculum consultant, JoAnne Buggey.
p. cm. — (Opposing viewpoints juniors)
Includes bibliographical references and index.
ISBN 1-56510-089-1 (acid-free paper)
1. Acid rain—Environmental aspects—Juvenile literature. 2. Acid rain—Law and legislation—Juvenile literature. 3. Critical thinking—Juvenile literature. [1. Acid rain. 2. Critical thinking.] I. Buggey, JoAnne. II. Title. III. Series.
TD195.44.A53 1994
363.73'86—dc20 93-14018
CIP
AC

Printed in the U.S.A.

CONTENTS

An Introduction to Opposing Viewpoints

When people disagree, it is hard to figure out who is right. You may decide one person is right just because the person is your friend or relative. But this is not a very good reason to agree or disagree with someone. It is better if you try to understand why these people disagree. On what main points do they differ? Read or listen to each person's argument carefully. Separate the facts and opinions that each person presents. Finally, decide which argument best matches what you think. This process, examining an argument without emotion, is part of what critical thinking is all about.

This is not easy. Many things make it hard to understand and form opinions. People's values, ages, and experiences all influence the way they think. This is why learning to read and think critically is an invaluable skill. Opposing Viewpoints Juniors books will help you learn and practice skills to improve your ability to read critically. By reading opposing views on an issue, you will become familiar with methods people use to attempt to convince you that their point of view is right. And you will learn to separate the authors' opinions from the facts they present.

Each Opposing Viewpoints Juniors book focuses on one critical thinking skill that will help you judge the views presented. Some of these skills are telling the difference between fact and opinion, recognizing propaganda techniques, and locating and analyzing the main idea. These skills will allow you to examine opposing viewpoints more easily. The viewpoints are placed in a running debate and are always placed with the pro view first.

Locating the Author's Main Idea

Authors include many ideas in their writing. But each sentence, each paragraph, and even each book they write should contain one main idea. For example, the main idea of this book is that acid rain is a point of debate.

Locating the author's main idea, whether it is within the sentence, paragraph, or entire piece of writing, is a basic reading skill. It is important because it allows readers to identify the theme of an author's writing. It also allows readers to understand the main point an author is trying to make about the theme.

In this Opposing Viewpoints Juniors book, you will be asked to analyze specific paragraphs to locate the main idea. Sometimes the main idea is placed at the beginning of the paragraph. Sometimes it is placed somewhere within the paragraph, or even at the end. For example:

> Oil producers have tapped most of the world's oil reserves for use by people in developed countries. They have also taken much of the world's natural gas and coal for use in power plants. If people are not forced to conserve soon, the world's natural resources will be gone in 100 years.

The main idea of this paragraph is placed at the end. It is that the world's resources are running out.

When you begin reading the paragraph, you might think the first sentence is the main idea. If it is, then the other sentences in the paragraph will support it in some way. They might explain the idea more specifically or give examples or reasons.

Read sentence two. Does it do any of these things? No. In fact, sentence two is very much like sentence one. Sentence two even says *also*, which suggests that the two sentences are giving two ideas about the same topic.

Now read the last sentence. It is a general statement about the world's resources, while the first two are specific examples of this. The last sentence in this paragraph is the topic sentence. The first two sentences support the last one.

If you outlined this paragraph, it would look like this:

1. If people are not forced to conserve soon, the world's natural resources will be gone in 100 years.
   A. (Example 1) Oil producers have tapped most of the world's oil reserves for use by people in developed countries.
   B. (Example 2) They have also taken much of the world's natural gas and coal for use in power plants.

Most paragraphs can be outlined in this way. By reading a paragraph carefully, you should be able to tell which sentence presents the main idea and which sentences explain or support it in some way. Outlining the paragraph may help you figure this out.

We asked two students to write one paragraph each in which they state their main ideas about acid rain. Examine the following viewpoints to locate the main ideas.

SAMPLE VIEWPOINT A *Cynthia:*

I think acid rain is getting worse.

I think acid rain is a bigger problem than ever. It has been killing trees and lakes all over the world, and it gets worse every year. It has attacked the forests of Scandinavia and Germany for the past thirty years. The forests of the eastern United States and Canada are also dying because of it. China, too, is starting to find trees and lakes damaged by acid rain. Even islands in the Pacific Ocean are showing acid rain damage. That means the effects of acid rain are spreading.

SAMPLE VIEWPOINT B *Jeffrey:*

I think acid rain is under control.

I think acid rain is under control. Ten years ago, a lot of people thought acid rain was like a spreading plague wiping out fish in lakes and trees in forests. But I read in the newspaper that a lot of scientists now say that's not true. Not much damage was done at all. And they're not even sure acid rain caused the damage. Scientists have learned a lot in the last ten years about acid rain. Now that scientists understand acid rain better, they can control it. We won't have to worry about it in the future.

ANALYZING THE SAMPLE VIEWPOINTS

Cynthia and Jeffrey have very different opinions about acid rain. Each presents one main idea in his or her viewpoint.

Cynthia:

MAIN IDEA

Acid rain is a bigger problem than ever.

Jeffrey:

MAIN IDEA

Acid rain is under control.

Both Cynthia and Jeffrey placed their main ideas at the beginning of their statements.

As you continue to read through the viewpoints in this book, remember to look for the main idea of each specified paragraph.

CHAPTER 1

PREFACE: Is Acid Rain a Serious Problem?

Acid rain is nothing new. The term *acid rain* was coined in the mid-nineteenth century by British scientist Robert Angus Smith. Smith used the term to describe the rain in his hometown of Manchester, England. Smith found that the rain had become acidic due to the air pollution from burning coal; that is, the rainwater had become sour and caused chemical changes in metals and other minerals. Today, scientists have found acid rain in many parts of the world. Actually, it would be more precise to say they have found that rain in many parts of the world has become *more* acidic. Natural compounds in the atmosphere, such as carbon dioxide, react with water to make all rain weakly acidic. So, rain, even "pure" rain, is naturally acidic, and always has been.

When people speak of acid rain, they are referring to rain that has become much more acidic than natural rainwater—sometimes as acidic as vinegar. When rain becomes this acidic, usually from mixing with air pollution, it can damage the environment. Soil, lakes, plants, buildings, monuments, machines, animals, and people can all be hurt by acid rain.

The debate about whether acid rain is a serious problem revolves around its ability to cause harm. Some experts claim that acid rain upsets ecosystems enough to kill living organisms. Fish die in acidic lakes; trees die in forests receiving acid rain; people have more lung diseases in areas with acid rain. These experts believe we should take steps to stop acid rain. If we do not stop it, they argue, it will get worse and its damage will increase.

Other experts do not believe acid rain is that serious. They admit that some lakes have become acidic and that some trees have died off in certain areas. But these experts hesitate to blame acid rain. Too many other factors must be considered, they say. They have noticed that not all areas where acid rain falls are affected in the same way. Sometimes areas where little acid rain falls suffer a lot of damage to trees and lakes. Meanwhile, areas that get a lot of acid rain do not suffer these effects. So, acid rain does not seem to be the cause of these changes. Therefore, these experts argue, stopping acid rain is not the answer. For them, acid rain is not that serious.

The following viewpoints debate the seriousness of the acid rain problem. Read each viewpoint carefully and read the questions in the margins. You will be asked to locate the author's main ideas.

VIEWPOINT 1 Acid rain is a serious problem

Editor's Note: This viewpoint argues that acid rain is a growing problem that needs attention. In its opinion, a serious acid rain problem exists and will only get worse if not faced. Read each paragraph carefully and answer the questions in the margin.

The main idea of this paragraph is stated in the first sentence.

Acid rain is one of the most serious environmental issues on earth. Scientists in the 1980s discovered that acid rain was damaging lakes, trees, and even buildings at an alarming rate in some parts of the world. Germany, Scandinavia, and the eastern United States were the hardest hit. Air pollutants were blamed, and environmentalists called for drastic reductions in air pollution.

The main idea of this paragraph is that some people believe acid rain is not a problem.

Today, however, some people no longer believe acid rain is to blame. They cite later studies that disagree with early reports. There is no acid rain disaster about to happen, they say, so why worry about acid rain?

What is the main idea of this paragraph? How do the sentences in the rest of the paragraph support the main idea?

But just because the acid rain problem may not be critical does not mean that a problem does not exist. Acid rain is still a problem because it continues to damage the environment. In fact, every day it threatens more regions of the world. According to a survey by the U.S. Environmental Protection Agency (EPA), nearly eleven thousand acres of lakes in New York's Adirondack Mountains have reached a critical state of acidification, and fish in two hundred lakes have disappeared. The EPA estimates that another nine thousand lakes in the northeastern United States are threatened. Some scientists and environmental organizations say these figures are too small. Even greater numbers of lakes and forests are threatened, they say.

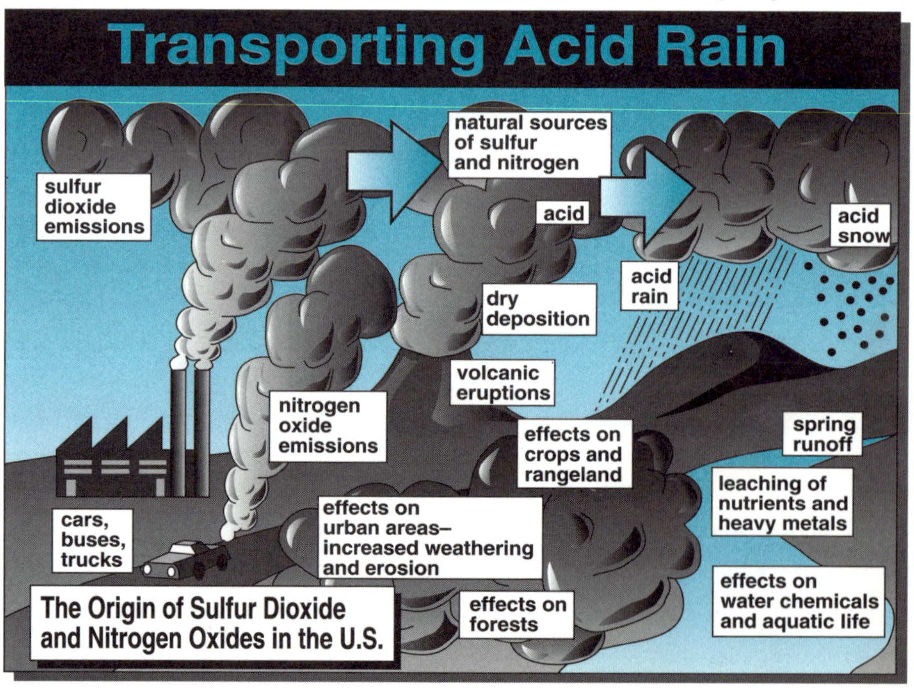

Increasing Acidity

Pure rainwater is naturally slightly acidic. Harmful acid rain is much more acidic—anywhere from pH4 to pH2 in some places—that is, as acidic as lemon juice. Such rain can lower the pH of lakes and soil to levels that kill fish and trees.

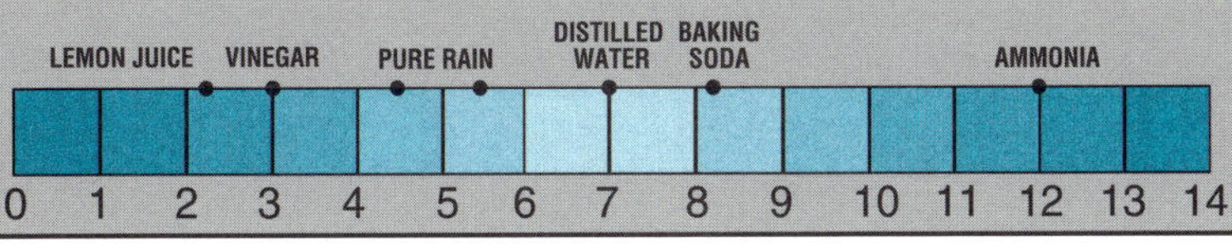

On the pH scale, the number 7 indicates a neutral solution. *Lower* numbers indicate increasing acidity. *Higher* numbers indicate increasing alkalinity. Each whole number represents a tenfold change: pH5 is ten times more acid than pH6. Therefore, pH3 is one thousand times more acid than pH6.

The number of damaged regions worldwide is also increasing. Acid rain has been found for decades in northern Europe and eastern North America. Now scientists are finding it in western North America, Japan, China, the former Soviet Union, and South America. It has even been found on the north slope of Alaska and on islands in the middle of the Pacific Ocean—places where there is no pollution. Winds have carried acid rain around the world. Such a spread is evidence that acid rain is becoming a global threat and must be stopped.

The main idea of this paragraph is that damage by acid rain is increasing. Does this paragraph support the main idea of the viewpoint? Why or why not?

Damage to human health is the most serious effect of acid rain. Environmental journalist Merritt Clifton writes that "acid rain is killing five to twenty times as many Americans as AIDS." As proof, Clifton cites studies done in the late 1980s by the Brookhaven National Laboratory of New York and the U.S. Office of Technology Assessment. According to these studies, between 50,000 and 200,000 deaths annually are caused by acid rain. A University of California study discovered that during a five-year period in the mid-eighties deaths from lung cancer rose 28 percent in Vermont. Vermont is located right in the middle of the so-called acid rain belt of the northeastern United States.

Which sentence in this paragraph contains the main idea? How does the rest of the paragraph help you determine this?

Acid rain creates another health hazard. It causes toxic metals to break loose from the naturally occurring chemical compounds in which they are harmless. The metals then seep into water and soil and end up in food supplies. These metals, such as aluminum, cadmium, and lead, enter people's bodies and cause cancer and other diseases.

Identify the topic sentence of this paragraph. How does this paragraph relate to the previous one?

Acid rain may not destroy the environment as quickly as some people once thought, but it will certainly cause suffering and mar a beautiful planet. People must recognize acid rain's seriousness as an environmental threat and take steps to stop it.

Where does the topic sentence appear in this final paragraph?

Still a problem

Why, according to the author, is acid rain a serious environmental issue? Why do some people think otherwise? Are you convinced by the author's arguments? Why or why not?

VIEWPOINT 2 Acid rain is not a serious problem

Editor's Note: This viewpoint argues that acid rain can no longer be considered a serious problem. Government studies have shown that acid rain will not be a significant environmental threat in the future. As you read each paragraph, locate its main idea.

The first paragraph is long and has a lot of information. Read through it carefully. The main idea is near the end. What is it? How do the quotations and the fable support it?

"The sky is not falling." With these words, James Mahoney, director of the National Acid Precipitation Assessment Program (NAPAP) in 1990, summed up the program's findings about acid rain. Mahoney's words refer to the fable of Chicken Little, who, after being hit on the head by a falling acorn, ran about hysterically announcing that the sky was falling. Chicken Little caused a great panic in the barnyard until someone actually looked up and saw the sky was still there and not falling after all. Mahoney used the fable to say that much time, money, and effort had been given to head off a disaster that simply was not happening. As Dr. Edward Krug, one of the NAPAP scientists, put it, acid rain was an environmental "nuisance, not a catastrophe." So why are we still hearing about it?

This paragraph is easier to outline. It makes a factual statement, then gives supporting statements to explain it. What is the paragraph's main idea?

In 1980, the Environmental Protection Agency (EPA) announced that acid rain was acidifying lakes in the northeastern United States. EPA stated that the number of acidic lakes had increased a hundred times since 1940. Scientists in the National Academy of Sciences predicted that that number would double by 1990. Virtually all of the lakes in the Northeast would be dead; that is, there would be no fish or other aquatic organisms living in them.

At about the same time, other scientists noticed that certain trees on mountaintops in Vermont and New York were dying. They blamed acid rain.

As a result of the outcries that resulted from these announcements, Congress commissioned NAPAP to determine the effects of acid rain. Ten years and $570 million later, NAPAP published the results of its studies in a six-thousand-page report. The results were surprising.

Which sentence contains the main idea of this paragraph? How does the paragraph support the viewpoint's main idea?

Acid rain, the studies found, had affected some lakes in the northeastern United States, just as scientists had first reported. But only about 10 percent were affected—hardly an epidemic. Some stands of red spruce trees on a few mountaintops in New Hampshire and New York had died, but not, as most environmentalists had claimed, from acid rain. Acid rain, the study found, was not a problem.

Krug concluded that the acid rain scare was the result of a wrong theory. The theory says that acid rain falling on normally alkaline (the opposite of acid) soil is neutralized and little or no harm is done. But in the northeastern United States, the soil is already acidic and will not

Effects of Current Levels of Acid Rain

neutralize the acid rain. The rain would just run off into lakes and streams and make them acidic. Thus, according to this theory, to stop lakes from becoming more acid, acid rain had to be stopped.

But Krug and his colleagues found evidence that contradicted this theory. For example, the Ohio River valley has the nation's most acidic rain, but its lakes and streams are not acidic. On the other hand, Florida has the greatest number of acidic lakes, but its rainfall is much less acidic than the rain in the Northeast. The scientists' studies revealed that the acidity of a lake or stream was determined by the kind of soil and vegetation around it and by how the land around the lake was used. Acid rain did not seem to matter.

What is the main idea of this paragraph? How does this paragraph continue the ideas presented in the previous one?

Such studies prove that trying to save lakes, streams, and forests from acidification by stopping acid rain will not work. Acid rain is an unimportant factor in the health of these natural resources and should not be thought of as a serious problem.

How does this paragraph support the viewpoint's main idea?

No big deal

The author states that the effects of acid rain on the environment are not serious. Having now read both viewpoints, do you agree or disagree? Why?

CRITICAL THINKING SKILL 1

Locating the Author's Main Idea

The following paragraphs each contain one main idea. Below each paragraph are three sentences, one of which best expresses the main idea of the paragraph. Read each paragraph carefully and circle the letter in front of the sentence that best expresses the main idea. More than one sentence may be related to the main idea, but you must choose the one that *best* expresses it. If you have difficulty locating a paragraph's main idea, first outline the paragraph as shown in the skill introduction on page 5.

Example: Air pollution from automobiles and power plants creates acid rain. The number of automobiles and power plants grows larger year by year. Acid rain will become a serious problem in the future if nothing is done to curb air pollution. This means restricting polluters such as cars and power plants.

a) The number of cars and power plants is increasing.
b) Acid rain damages forests and lakes.
c) Acid rain may become a serious problem.

The answer is *c* because it best expresses the point the author wants to make in the paragraph. Sentence *a* supports the main idea. Sentence *b* is not discussed in the paragraph.

1. Acid lakes are a natural occurrence. They are caused by a combination of factors. The acidity of the soil, how the surrounding land is used, and the kinds of plants growing nearby all affect acidity in lakes. Human-caused acid rain will not measurably affect, let alone cause, the acidity of most lakes.

   a) Many factors cause the acidification of lakes.
   b) Acidification of lakes is a natural occurrence.
   c) Acid rain affects lakes.

2. Growing industrialization in many areas of the world contributes to the acid rain problem. Acid rain has now been found even in areas far from industrial centers. It is carried there by wind currents high in the atmosphere. Soon, there will be nowhere on earth untouched by acid rain's damaging influence.

   a) Industrialization causes acid rain.
   b) Atmospheric winds carry acid rain around the world.
   c) The whole earth is threatened by acid rain.

3. Dr. Edward Krug is a soil scientist who worked on the U.S. government's study of acid rain. Krug found that acidification was mostly a natural process and not caused by acid rain. His findings contradicted the prevailing theory that acid rain was killing fish and trees.

   a) Krug's studies disagreed with accepted acid rain theory.
   b) Krug is a soil scientist.
   c) Acid rain does not damage lakes and forests.

CHAPTER 2

PREFACE: Is Acid Rain Harming Lakes and Streams?

A lake, pond, or stream is an ecosystem; that is, it is an environment in which different kinds of animals, plants, and minerals all coexist. Even a small pond is a complex ecosystem. It has certain kinds of plants that thrive in the particular kind of soil in and around the pond. Certain animals that live off those types of plants will live in or near the pond. And other animals that live off those animals will also thrive there. If anything changes in the ecosystem it affects all the species that make up the system. If the acidity of the soil changes, for example, the plants that grow in the soil have to adapt to the change. If the change is too great and the plants cannot adapt, they die. Without them, the animals that depend on them for life have to adapt or die, too.

When scientists in the 1970s noticed that many lakes in northeastern New York state were losing their fish due to acidification, they blamed acid rain. The National Academy of Sciences predicted that by the year 2000 the number of acidified lakes would increase a hundredfold if acid rain was not stopped. A very gloomy future for the environment was forecast, and the nation was scared.

Later government-sponsored studies revealed that the alarm had been a false one. Acid rain was not a critical problem. Many people breathed a sigh of relief and turned their attention to other, more pressing problems. But the debate over the effects of acid rain on lakes and streams did not die.

Environmentalists continue to argue that acid rain is damaging lakes and other aquatic ecosystems. The Alliance for Acid Rain Control, for example, claims that 25 percent of smaller lakes in New York state are acidic because of acid rain. Others note that though the problem might not be as bad as was once thought, it is still a problem and should be remedied. Why wait until a problem becomes a crisis before fixing it? they ask.

Those who disagree with this point of view believe that most acidification of lakes and streams is a natural process. To spend millions of dollars to stop it by stopping acid rain would be to waste money, they argue.

The following two viewpoints debate whether acid rain is harming lakes and streams. Be sure to answer the questions in the margin.

VIEWPOINT 3 Acid rain is harming lakes and streams

Editor's Note: The following viewpoint argues that acid rain is harming lakes and streams. Studies that have shown the damage to be less than predicted should not be taken as proof that there is no problem at all. Canada's acid rain problem provides ample proof that acid rain indeed harms lakes and streams.

The first paragraph functions simply to introduce the viewpoint. What is its main idea?

Acidified lakes and streams were the first heralds of the acid rain crisis. Scientists in the 1970s noticed changes in these delicately balanced ecosystems. They did not like what they saw. The scientists sounded an alarm.

What is the main idea of this paragraph? How do the other sentences support it?

Scientists in the northeastern United States first noted that fish and other water creatures were disappearing from many lakes. They also found that the water in the lakes was getting clearer. To the untrained eye, such a sight might seem like an improvement. Clearer water means cleaner, more inviting water—if you are a human being looking to go swimming. But to fish, frogs, worms, and water plants, it spells trouble. A clean lake means an empty lake, one in which living plants and animals are not stirring things up.

When the scientists tested the lakes, they found that the water had become too acidic for many creatures to live in. They blamed acid rain for raising the acidity of the lakes. And they blamed human-made air pollution for causing acid rain.

The scientists predicted that if acid rain kept falling into the lakes and streams, thousands of bodies of water would die. This alarmed environmentalists, who then pressured the government to force polluters to reduce pollution. But because that action would cost a lot of money, the government wanted to study the problem first.

Those studies showed that many lakes and streams were becoming more acidic. But not all the evidence pointed to acid rain as the cause. Many government officials and business people then laughed at the environmentalists' fuss about acid rain. They did not want to spend the billions of dollars that environmentalists said would be necessary to stop acid rain. So they began to say that the acid rain problem was an illusion, that there was no problem.

But there is a problem. Lakes and streams *are* dying because of acid rain. Maybe not hundreds of thousands of lakes and streams. Maybe only (only!) tens of thousands. According to Ned Helme and Chris Neme of the Alliance for Acid Rain Control, 15 percent of medium and large lakes and 25 percent of small lakes in New York's Adirondack Mountains are acidic. Helme and Neme blame acid rain.

It is wise to remember also that the United States is not the only place endangered by acid rain. Canada appears to suffer much more from the effects of acid rain than does the United States. Most of Canada's acid rain is caused by American polluters. Canadian studies show that much damage is being caused by acid rain. More than a million square miles of eastern Canada is vulnerable to acid rain damage. Studies conducted in the United States found that about

This chart shows that not all fish, or their food insects can tolerate the same amount of acid. Fish like trout, bass, and perch are affected at different pH levels. Which type of fish are the most sensitive to acid? Generally, the young of most species are more sensitive than adults. Frogs may tolerate high levels of acidity, but if they eat insects like the mayfly, they may be affected because part of their food supply may disappear.

11,000 U.S. lakes were vulnerable to acid rain. Canada has 300,000 vulnerable lakes. Half of these are now suffering damage, and 14,000 lakes are already dead. Dead U.S. lakes numbered 1,100.

The Canadian government is convinced that acid rain *is* damaging Canada's lakes and streams. Pollution reduction in Canada has helped to identify and prevent acid rain damage in nearby lakes. Such results clearly indicate that acid rain harms lakes.

Where is the main idea of the last paragraph expressed? Does this sentence support the main idea of the viewpoint? How?

The Life and Death of a Lake Ecosystem

To support normal aquatic life, a lake habitat must have a balanced food chain and allow for normal reproductive cycles. When acid is introduced to the ecosystem, either through acidic rainfall or runoff from acidic snowpack, the influx of acid may alter the lake's overall pH, creating a more acidic environment. Also, acidic runoff can release heavy metals—highly toxic for many aquatic species—from rocks and soil. Sensitive organisms may be wiped out and resistant life forms are weakened as their food supplies are affected.

- Fish prey on insects and provide food for certain birds.
- Fish eggs hatch in the normal way.
- Insects provide food for fish and amphibians.
- Hunting birds, such as the osprey, live on fish.
- Frogs feed on insects.
- Trees are supported by local soil.

Source: *National Geographic*

- Acid influx and heavy metal poisoning can kill sensitive fish, such as rainbow trout and salmon.
- Fish eggs can be destroyed or deformed by environmental changes.
- Certain insects, such as the whirligig beetle, reproduce uncontrolled. They are resistant to acidification and their main predators are fewer.
- Birds must seek fish in other lakes.
- Frogs die in heavily acidic habitats.
- Acidic soil can damage or even kill certain trees.

How many damaged lakes are too many?

The author makes the point that a problem is still a problem even though it may be less critical than people once thought. Does it seem wise to ignore a problem until it grows into a dangerous crisis? What do you think?

VIEWPOINT 4

Acid rain is not harming lakes and streams

Editor's Note: The following viewpoint argues that acid rain is not harming lakes and streams. Rather, lakes and streams are acidic because of the soil and vegetation that surrounds them.

Which sentence in this paragraph states the paragraph's main idea?

In 1980 the Environmental Protection Agency (EPA) declared that lakes in the northeastern United States had become one hundred times more acidic than they had been forty years earlier. The National Academy of Sciences even predicted that that number would double by 1990. As a result of this apparent environmental crisis, Congress established the National Acid Precipitation Assessment Program (NAPAP), to be headed by the EPA.

What is the topic sentence of this paragraph? How does this paragraph support the main idea of the viewpoint?

For years, environmentalists tried to pressure the government into spending millions of dollars to reduce acid rain. But the government first wanted to get the results of NAPAP's studies. When the data began to come in, it was not what most people expected. The EPA's initial declaration and the National Academy of Sciences' dire prediction proved to be wrong. Acid rain was not acidifying lakes!

How does this paragraph support the main argument of the viewpoint? Does it logically follow the previous paragraph? Why or why not?

NAPAP studied, among other areas, the lakes in northeastern New York's Adirondack Mountains. This area was thought to be the most threatened by acid rain damage. It was the acidity of these lakes that, according to the EPA, had increased a hundredfold in forty years. The studies found that the lakes were no more acidic now than they were before the Industrial Revolution in the early nineteenth century. (Environmentalists blamed industrial pollution for the increase of acidity in lakes.) And, contrary to the prediction of the National Academy of Sciences in 1980, the studies found that acidity in those

"SAVE $3-BILLION A YEAR IN FEDERAL CLEAN-AIR REGULATION BY SIMPLY CLEANING ACIDIFIED LAKES WITH LIME? WHAT, AND LEAVE ALL THESE EPA BUREAUCRATS WITHOUT A JOB?!"

lakes had not doubled between 1980 and 1990; in fact, the acidity was unchanged.

Samples of the soil beneath the lakes were taken for study. These samples told an interesting story about the lakes' acidity. The samples showed evidence of increased acidity in the recent past. But they also revealed that many of the lakes had been acidic and without fish even before European settlers arrived in the area.

The main idea of this paragraph is expressed neither in the first nor the last sentence. Which is the topic sentence?

In other words, the lakes were naturally acidic. They had become more alkaline (the opposite of acidic) around the turn of the twentieth century. At that time, large-scale logging operations cleared away most of the vegetation in the area. The vegetation was acidic. In its absence, the soil became more alkaline and so did the lake water. These conditions were more favorable for trout and other fish, which then thrived in the lakes for decades.

How does this paragraph relate to the previous one? What is its main idea?

Slowly the forests and undergrowth returned, however. The soil and lakes again became acidic, and the fish disappeared. All of this was a natural process unrelated to acid rain. As NAPAP scientists Dr. Edward Krug and Charles Frink concluded in a controversial 1983 report titled "Acid Rain on Acid Soil," plants and acid soil, not acid rain, caused acid lakes. Krug and Frink's study included lakes in the unspoiled New Zealand environment. No acid rain fell there, but the lakes were very acidic. The pair also noted that Florida has many acidic lakes but little acid rain. The Ohio River valley, on the other hand, has the highest concentration of acid rain but almost no acid lakes.

This paragraph continues the thought of the preceding paragraph. Its main idea is nearly identical. Where is it expressed in the paragraph?

The evidence of these studies contradicts the claim that acid rain is harming lakes and streams. Some lakes and streams are naturally acidic; others are not. Acid rain or its absence does not seem to matter one way or another.

The concluding paragraph sums up the viewpoint's main idea. Which sentence best expresses it?

Acidic Lakes Past and Present

Studies show that many lakes are no more acidic now than in the mid-1800s.

	pH in 1850	pH in 1986-88	pH change	Actual
Adirondack acidic lakes				
Under pH5.5	4.95	4.78	-.17	none
Under pH5.0	4.78	4.63	-.15	none
Florida acidic lakes				
Under pH5.5	5.11	4.81	-.30	none
Under pH5.0	5.00	4.65	-.35	more acidic

Source: Environmental Protection Agency.

No harm done

The author argues that studies disprove the theory that acid rain causes lakes and streams to become acid. Does his argument persuade you to agree with him? Why or why not?

CRITICAL THINKING SKILL 2
Selecting the Main Idea

The following paragraphs are taken from the viewpoints in this chapter. Each paragraph contains one main idea. Below each paragraph are spaces to outline the paragraph by stating the main idea along with the supporting ideas. Read carefully and outline each paragraph below. Compare your answers with those of your classmates.

Example: Acidified lakes and streams were the first heralds of the acid rain crisis. Scientists in the 1970s noticed changes in these delicately balanced ecosystems. They did not like what they saw.

Main idea: Acidified lakes and streams heralded the acid rain crisis.

Supporting ideas: Scientists noticed changes in lakes and streams.
Scientists did not like the changes they saw.

1. But there is a problem. Lakes and streams *are* dying because of acid rain. Maybe not hundreds of thousands of lakes and streams. Maybe only (only!) tens of thousands. According to Ned Helme and Chris Neme of the Alliance for Acid Rain Control, 15 percent of medium and large lakes and 25 percent of small lakes in New York's Adirondack Mountains are acidic. Helme and Neme blame acid rain.

Main idea:

Supporting ideas:

2. For years, environmentalists tried to pressure the government into spending millions of dollars to reduce acid rain. But the government first wanted to get the results of NAPAP's studies. When the data began to come in, it was not what most people expected. The EPA's initial declaration and the National Academy of Sciences' dire prediction proved to be wrong. Acid rain was not acidifying lakes!

Main idea:

Supporting ideas:

3. The evidence of these studies contradicts the claim that acid rain is harming lakes and streams. Some lakes and streams are naturally acidic; others are not. Acid rain or its absence does not seem to matter one way or another.

Main idea:

Supporting ideas:

CHAPTER 3

PREFACE: Does Acid Rain Endanger Forests?

In the early 1980s, scientists reported that the red spruce blanketing the top of Camel's Hump Mountain in Vermont had turned brown and were dying. They concluded that acid rain was the cause. Similar damage had been reported in forests in Germany, Scandinavia, and Canada. University of Vermont botanist Hubert Vogelmann described the situation: "Twenty years ago the evergreen forests on the slopes of Camel's Hump . . . were deep green and dense. . . . Today the red spruces are dead or dying and some firs look sick. Gray skeletons of trees . . . are everywhere in the forest."

The U.S. government funded studies to follow up on this report. These studies found that acid rain's role in forest decline was unclear. There was some evidence that acid rain adversely affected forests. But there was just as much evidence to the contrary. This is because many factors affect a forest ecosystem, many more than are involved with a lake or stream. It is extremely difficult to pinpoint one factor as a cause of decline. Thus, the government-sponsored National Acid Precipitation Assessment Program (NAPAP) finally concluded that there "is no evidence of an overall or pervasive [widespread] decline of forests in the United States and Canada due to acidic deposition [precipitation]."

Environmentalists do not agree. They claim that, due to this report, the 1990 Clean Air Act reduces acid rain only enough to save lakes from acidification but not enough to stop forest decline. Environmental writer John Flynn reported in the *Amicus Journal* that "reductions of sulfur and nitrogen emissions required by the act will not be nearly sufficient to protect sensitive forest systems from continuing decline and mortality." Many believe, therefore, that acid rain still endangers forests.

Among those who believe acid rain presents no danger to forests is former NAPAP director J. Laurence Kulp. Kulp sees acid rain as more of a benefit than a harm to forests. It is beneficial, he says, "because the nitric acid-nitrate component of acid rain acts as a fertilizer and enhances tree growth in general."

Does acid ran endanger forests? The viewpoints in this chapter debate this issue. Remember to answer the margin questions as you read.

VIEWPOINT 5 Acid rain endangers forests

Editor's Note: Despite the government's report to the contrary, acid rain's threat to forests is very real, this viewpoint argues. It cites studies by scientists who believe acid rain is damaging forests.

Which sentence expresses the main idea of this paragraph? How do the other sentences support the main idea?

Forests, like lakes and streams, are threatened by acid rain. The forests of red spruce atop Vermont's Green Mountains and in the Adirondacks of northern New York state, for example, are visibly declining. The trees show evidence of yellowing needles at their tops and on their side branches. The size of their roots has decreased, and they die too early. Other cases abound in America's South and West and in Canada and Europe.

Is this paragraph's topic sentence the first or the last? Why?

The decline of trees in these areas was one of the first indications to scientists in the 1980s that something was seriously wrong with the environment. The areas are downwind of the heavily industrialized areas of Ohio and Pennsylvania. Scientists logically concluded that acid rain was responsible for the trees' sudden sickness.

Does this paragraph support the viewpoint's main idea? Why or why not?

In 1984, the federal government's National Acid Precipitation Assessment Program (NAPAP) conducted studies on those forests. It found no evidence of "a general or unusual decline of forests in the United States and Canada due to acidic deposition or any other stress factor." They attributed the deaths of the trees to insects or severe weather.

Effects of Acid Rain on Trees

Type of Tree	Region Found	Effect
Norway Spruce; Beech	Germany; Central Europe	Death; thin tops; new growth retarded; yellowing of leaves
Spruce; Fir	Eastern United States	Yellowing; growth rate decrease; death
Southern Pines	Southeastern United States	Growth rate reduced
Sugar Maples	Northeastern United States Southeastern Canada	Dead tops; reduced syrup production
Ponderosa and Jeffrey Pines	Central and Southern California	Death; reduced growth rate
All forests	Throughout Earth's Temperate Zone	Soil acidification; leaching nutrients from soils

Source: NAPAP

James MacKenzie and Mohamed T. El-Ashry of Washington, D.C.'s World Resources Institute disagree. Their research contradicts NAPAP's findings. They contend that NAPAP's conclusions "are, at least, misleading." The two scientists maintain that "the scientific evidence is unambiguous [unmistakably clear] that ozone pollution and acid deposition—both together and separately—are contributing to the deaths of trees in the South, East and West." They argue that the stress of acid rain and other pollutants weakens trees so much that they fall prey to insects, diseases, and extreme weather conditions that they would normally be able to withstand.

How does this paragraph relate to the preceding one? What is the main idea of this paragraph?

It is difficult to measure acid rain's impact, MacKenzie and El-Ashry admit, because the rain does not usually kill the trees outright. It weakens them so that they cannot fight off natural enemies. The trees' plight can be compared to that of humans infected by the AIDS virus. AIDS victims do not die from AIDS. They die from pneumonia or some other disease. This is because the virus has weakened their immune systems too much to withstand infections that healthy people survive.

Does the AIDS analogy support this paragraph's main idea? Why or why not?

The effects of acid rain on forests is more complicated than the effects on lakes and streams. A forest is a more complex ecosystem than a lake or stream. Soil and groundwater conditions, parasites and predators, weather, and locality all play a part in the lives of forests. Scientists must account for acid rain's effect on any and all of these factors before they can make an accurate assessment of its effects on forests.

Some researchers accuse NAPAP of wanting too easy an answer to the acid rain problem. They charge that the organization overlooked or even denied some evidence. For example, Robert Bruck of North Carolina State University did a study for NAPAP on acid rain's effects on dying trees in his state's forests. He concluded that acid rain made the soil so acidic that it killed the trees. NAPAP ignored his evidence, he says, and reported that insects caused the trees' deaths.

Acid rain does indeed endanger forests. To ignore the evidence and continue in the unrealistic belief that acid rain is no problem for forests will prove disastrous.

"I don't care what the government says—That acid rain makes me hallucinate!"

A complicated problem

Why is it difficult to determine acid rain's impact on forests, according to this viewpoint? Do you agree with the viewpoint's argument that NAPAP's conclusions about acid rain and forests are doubtful? Why or why not?

VIEWPOINT 6 Acid rain does not endanger forests

Editor's Note: This viewpoint contends that the evidence showing that acid rain harms forests is questionable. People who believe acid rain endangers forests are not taking into account all the factors affecting forest ecology.

How do the sentences of this paragraph support its main idea? What sentence expresses the paragraph's main idea?

"The consensus so far after ten years of research nationwide is that there is no documentation [evidence] for believing that acid rain is causing widespread harm to North America's commercial forests." This is the opinion of Ivan Fernandez, an associate professor of soil sciences at the University of Maine. Fernandez and his colleagues are studying the effects of acid rain on forest trees and soil. They have found higher levels of sulfur and nitrogen in forest soil after acid rain falls upon it. But these levels do not harm trees, says Fernandez. In fact, the higher levels of nitrogen, which is an important plant fertilizer, is good for the trees. Though the trees in Fernandez's research show no harm from acid rain, he concedes that trees on the tops of mountains may find it stressful. Acid rain has been blamed by some scientists for tree damage atop mountains in Vermont and New York.

Locate the main idea of this paragraph. How does the last sentence of the preceding paragraph set up the transition into this idea?

Yale University forestry expert Tom Siccama, however, does not believe acid rain caused the deaths of red spruce trees atop Vermont's Green Mountains. Trees on these mountains, particularly Camel's Hump Mountain, died off in the early 1980s. Scientists blamed acid rain. That story, spread by the media, started the uproar over acid rain. But according to Siccama, the trees died too suddenly for acid rain to be the cause. Soil made too acidic from acid rain could not have been the problem, Siccama states, because new seedlings were springing up all over the mountaintop. In fact, the overall health of the forest was improving, not getting worse. What caused the trees' deaths? According to Siccama, a severe drought and an extremely cold winter back in 1962 weakened the trees so much they died prematurely. Trees growing up after 1962 were healthy and did not die.

What is the main idea of this paragraph? How does the first sentence of the paragraph act as a transition from the previous paragraph?

J. Laurence Kulp, former director of the National Acid Precipitation Assessment Program (NAPAP), agrees with Siccama. Kulp reports that NAPAP research found that a similar decline in red spruce in the area occurred from 1870 to 1880. That decline happened when there was very little acid rain. So acid rain could not have caused it. It was due to the unusually severe winters during that period, says Kulp.

Too many factors are involved to blame just one for any decline in tree health. Professor Paul Mannion of the State University of New York concludes:

If one recognizes the complex array of factors that can contribute to the decline of trees, it is difficult to accept the hypothesis that air pollutants [the cause of acid rain] are the basis of our tree decline problems today.

To simply state, then, as some scientists do, that forests are harmed because acidic rain is being added to already acidic soil is to make an unscientific conclusion. Much more study on the interaction of all the factors involved must be done before an accurate conclusion can be drawn.

What is the main idea of the last paragraph? How does it support the main idea of the viewpoint?

Acid Rain Fertilizer

U.S. FORESTS—GROWING STOCK VOLUME

(In millions of cubic feet)

		1952	1977	1987	% CHANGE
United States	Total:	610	725	755	23.8
	Softwoods:	430	465	450	4.7
North	Total:	106	166	190	79.2
	Softwoods:	28	46	49	75.0
Northeast	Total:	61	100	109	78.7
	Softwoods:	18	31	31	72.2
New England	Total:	24	42	44	83.3
	Softwoods:	14	23	22	57.1

Source: U.S. Forest Service, U.S. Statistical Abstract.

Acid rain is not destroying forests. On the contrary, it helps fertilize the soil with nitrates. Agriculturalists have known about this "free fertilizer" for a hundred years and have welcomed it. The chart shows that the volume of trees in the United States is increasing, not decreasing—especially in the Northeast, where the rain is most acidic. How then can acid rain be wiping out forests, as some environmentalists have claimed?

Other factors involved

This viewpoint cites evidence and quotes experts regarding other reasons besides acid rain for the decline of forests in some places. Now that you have read both viewpoints on forests and acid rain, do you think sickly forests would regain their health if acid rain were stopped? Why or why not?

CRITICAL THINKING SKILL 3
Developing the Main Idea

Listed below are ten statements. Each is related to information you have read in this book. Choose one and make it the main idea of a paragraph you write. Remember as you write that your goal is 1) to write sentences that are related to the main idea, and 2) to add specific information, like facts or statistics, to develop the main idea.

Example: Topic idea: Air pollution causes acid rain.

Acid rain is caused by air pollution. The principal pollutants that cause acid rain are nitrogen oxides and sulfur dioxide. These chemicals are released into the air by car engines and coal-burning power plants. There they mix with water vapor. A chemical reaction occurs. The products of that reaction are nitric acid and sulfuric acid. The water vapor and acid then fall to earth as acid rain.

Main Ideas:

1. Acid rain is bad for people's health.
2. Acid rain now threatens the whole earth.
3. Red spruce on Camel's Hump Mountain died from acid rain.
4. Many factors contribute to the acidification of lakes.
5. Acid rain does not kill trees outright.
6. NAPAP's report ignored some important data.
7. Rainwater is naturally acidic.
8. Acid rain is a big problem in Canada.
9. Scientists disagree about acid rain's effects on forests.
10. Acid rain is good for trees.

CHAPTER 4

PREFACE: Are Laws Necessary to Reduce Acid Rain?

In the fall of 1990, Congress passed the Clean Air Act Amendments. These amendments to the 1970 Clean Air Act included a program expressly for reducing acid rain. The law called for the reduction of specific air pollutants that cause acid rain, primarily sulfur dioxide and the oxides of nitrogen. The law stated that sulfur dioxide emissions must be reduced by 50 percent and nitrogen oxides emissions must be reduced by 15-25 percent by the year 2003. The cost to do this was estimated to be about $100 billion. People were divided about whether such a law was really necessary.

Environmentalists and liberals argued that the Clean Air Act needed to deal directly with acid rain. Previous environmental laws had not solved the problem of acid rain, they maintained. They argued that industrial polluters would never reduce their pollution unless forced to do so by law. A new law was necessary if acid rain is to be reduced, the liberals and environmentalists insisted.

Conservatives and many economists argued that such a law was totally unnecessary. They cited studies that show acid rain has little to do with the acidification of lakes and forests. It made no sense, they argued, to pass a law reducing acid rain—especially if it would cost billions of dollars to put into effect. They saw such a law as a waste of money and a threat to the nation's economy.

The law was passed, but the debate continues. The following viewpoints are examples of that debate. In this final chapter, no margin questions are offered to help you locate the author's main ideas. Read the viewpoints carefully and use the skills you have learned to locate each paragraph's main idea.

VIEWPOINT 7 Laws are necessary to reduce acid rain

Editor's Note: Legislation has always been the most effective means of controlling pollution, this viewpoint argues. Legislation is also the way the public ensures that its will is done. Read the viewpoint carefully. Locate the main idea in each paragraph and answer the questions in the focus box at the end of the viewpoint.

Laws reducing air pollution are necessary to reduce acid rain. Historically, legislation has proved to be the most effective way of reducing pollution. The first Clean Air Act, passed in 1963, provided for scientific study of pollution but did not restrict polluters or pollutants. In 1970, Congress amended the Clean Air Act, setting limits for pollutants and placing restrictions on polluters. Congress established the Environmental Protection Agency (EPA) to enforce the new limits and to monitor how well corporations obeyed the new laws. As a result, pollution was significantly reduced. For example, the EPA forced auto manufacturers to install emissions controls on automobiles. The automakers strongly resisted. They complained that consumers would not pay the extra cost of these devices. They predicted job losses and economic disaster. Their dire predictions did not come true, however, and pollution from automobile exhaust has been reduced by almost 100 percent.

By the 1980s, the damage from acid rain had become apparent. The chief cause was sulfur dioxide spewed into the air by coal-burning electric power plants. In the 1970s the EPA had set less-strict controls on power plants. It did so because electrical energy is important to the nation's economy. The EPA did not want to cause economic problems by forcing the plants to spend millions all at once to install pollution-reducing equipment. The EPA gave the owners of power plants several years to install such equipment. Again and again, the deadlines for compliance were extended. In other words, little change occurred. Edward Edelson, author of the book *Clean Air*, reports that the steadily improving air quality in the United States started to decline again. Something had to be done.

A new, stricter Clean Air Act was demanded. Years passed as Congress and the presidential administration, coached by environmental and corporate lobbyists, hammered out new laws. Finally, in 1990, changes in the Clean Air Act were passed into law. These changes included controls specifically aimed at reducing acid rain.

Conservatives and economists denounced the $100 billion price tag that implementing the new laws over the next decade would cost. Environmentalists considered it a step forward, but by no means did they believe the law accomplished everything that needed to be done about air pollution.

The public wants a clean environment, and it wants to stop acid rain. Passing laws ensures that the will of the people is done, not just the will of big business. Consumer watchdogs David Bollier and Joan Claybrook expressed this in their book, *Freedom from Harm: The Civilizing Influence of Health, Safety, and Environmental Regulation:*

> Regulation is an imperfect but necessary attempt to democratize corporate-owned science and technology. It injects public concerns into otherwise "private" decisions and investments. It helps the public understand and interpret technological information that otherwise is jealously guarded by the corporate scientific elite.

Without laws, industrial polluters would pursue their own interests—making profits—while ignoring the public's best interests.

The will of the people

The viewpoint states that legislation may not be the most perfect way of controlling acid rain, but it has an important function in a democracy. What is that function? Why is it important? How does this idea support the viewpoint's main idea?

VIEWPOINT 8 Laws are not necessary to reduce acid rain

Editor's Note: This viewpoint argues that the acid rain law passed in 1990 is unnecessary and will only result in a waste of money. The law is misguided, the viewpoint says, because acid rain is not the cause of acidification of lakes and forests. As you read, remember to locate the main idea of each paragraph.

In 1990, Congress amended the Clean Air Act. The act now has a section specifically on acid rain. It requires that the owners of manufacturing plants and electric power plants install emissions-reducing equipment. The goal is to reduce sulfur dioxide (SO_2) pollution by ten million tons a year by early in the next decade. That is about half of the present amount of SO_2 now expelled into the air. Half of the reduction is scheduled to be achieved by 1996. The total cost is estimated to be about $100 billion.

To many people, this program sounds expensive but worthwhile since it will greatly reduce the damage from acid rain. But it is completely unnecessary. It will waste billions of dollars and will not affect the acidification of lakes, streams, and forests. J. Laurence Kulp, the former director of the National Acid Precipitation Assessment Program (NAPAP), has stated, "The crash effort to reduce emissions . . . is estimated to cost on the order of $100 billion over the next 20 years and will not permanently solve the problem."

Kulp's successor as the head of NAPAP was James Mahoney. In the fall of 1990, Mahoney made the following statement to the U.S. Senate Environment Committee:

> If there were no change in acid deposition from current levels, we would expect that in the Northeast the chemical status of the lakes would stay approximately the same over the next 50 years.

He was asked what would happen to lakes if the levels of acid rain doubled, as many had predicted they would. His answer: "Not much."

Despite the reports of these experts, Congress passed the amendments to the Clean Air Act. It seems Congress was just trying to please environmentalists. It ignored the findings of NAPAP and now has unnecessarily burdened the American people with a huge expense that will give little benefit in return. As Kulp testified, "Compared with the $100 billion cost, it is hard to identify $1 billion in benefits."

According to Kulp, this costly program to hurry to reduce acid rain was based on two false beliefs: (1) The current level of acid rain has produced an ecological "crisis," and (2) immediate action is necessary because the situation is getting rapidly worse. The $500 million NAPAP study showed that acid rain had a very small effect on the environment. Some lakes did become acidified—but only about seventy-five, according to journalist Warren Brookes, a long-time opponent of the expensive acid rain amendment. Kulp concludes, "No worsening of the situation is expected in the next half century even without additional regulation. There is no reason for a 'crash' strategy such as is built into the 1990 Clean Air Act."

Passing laws as a way to bring about a desired change is usually an effective action. In the case of acid rain it will be fruitless. The laws will cost huge amounts of money and bring about nothing because acid rain is not the cause of most acidification. America is going to pay a high price for a "cure" for a nonexistent "disease." An expensive lesson indeed!

Henry Payne reprinted by permission of UFS, Inc.

An unnecessary expense

The viewpoint argues that a law calling for expensive measures to reduce acid rain is unnecessary. Do you agree? Why or why not? Do the main ideas of each paragraph support the main idea of the viewpoint? Why or why not?

CRITICAL THINKING SKILL 4

Locating the Main Idea in Editorial Cartoons

In this book you have seen cartoons that illustrate the ideas in the viewpoints. Cartoonists use humor to present their opinions on an issue. Like other writers, they express a main idea in their cartoons. Sometimes the main idea is obvious. At other times, a reader must look for clues in the words and drawings to determine the main idea.

Helpful questions to ask when reading editorial, or "opinion," cartoons are: What or who do the pictures represent? What does the caption (the words) say about the pictures? What hints about the meaning are given by the kinds of things pictured and how are they drawn or arranged? What is the main idea, the point, of the cartoon?

Look at the cartoon below. Answer the above questions about the cartoon. The following additional questions may be helpful in understanding the main idea of this cartoon: Which one of the six pictures does not belong with the others? Which things do environmentalists try to save? Who do environmentalists usually fight against to save natural resources? (Hint: *expendable* means that something can be used up and replaced so it does not need saving.)

What is the main idea of this cartoon? Does the cartoonist side with or against environmentalists? How can you tell?

Asay, by permission of the *Colorado Springs Gazette Telegraph.*

For further practice, look at the cartoons elsewhere in this book. Try to determine what the cartoons' main ideas are.

You may also want to try drawing a cartoon yourself that expresses an idea about acid rain. You may wish to use one of the ideas from Critical Thinking Skill number 3. Show the cartoon to someone else and see if they can figure out the main idea it expresses.

FOR FURTHER READING

The author recommends the following books and periodicals for further research on the topic. Check the works consulted list that follows for further suggestions.

Robert Anderson, *Forests*, San Diego: Greenhaven Press, 1992.

Robert Anderson, *Pollution*. San Diego: Greenhaven Press, 1992.

A. Joseph Dowd, "Science and Environmental Politics," *Vital Speeches of the Day*, July 15, 1991.

Edward Edelson, *Clean Air*. New York: Chelsea House, 1992.

Environment, "Overview," July 1989.

Eileen Lucas, *Acid Rain*. Chicago: Childrens Press, 1991.

Christina G. Miller and Louise A. Berry, *Acid Rain: A Sourcebook for Young People*. New York: Julian Messner, 1986.

Don Nardo, *Ozone*. San Diego: Lucent Books, 1991.

Gail B. Stewart, *Acid Rain*. San Diego: Lucent Books, 1990.

Maria Talen, *Ocean Pollution*. San Diego: Lucent Books, 1991.

WORKS CONSULTED

The following books and periodicals were used in the compilation of this book.

Terry L. Anderson and Donald R. Leal, *Free Market Environmentalism*. Boulder, CO: Westview Press, 1991. Describes how pollution and other environmental problems are best solved by use of free-market principles because such a method makes it profitable to prevent and clean up pollution.

William Anderson, "Acid Test," *Reason*, January 1992. Describes how EPA soil scientist Edward Krug was fired and his work attacked after he announced his findings that acid rain was not an environmental disaster.

David Bollier and Joan Claybrook, *Freedom from Harm: The Civilizing Influence of Health, Safety, and Environmental Regulation*. Washington, DC: Public Citizen and Democracy Project, 1986. Describes how and why government environmental and safety regulations work for the public's good.

Warren T. Brookes, "Congress Ignores the Facts on Acid Rain," *Conservative Chronicle*, April 11, 1990. Argues that passage of the 1990 Clean Air Act occurred despite scientific evidence proving that the act's proposed high costs are unnecessary.

Derek Burney, "Canada and the United States in a Global Context," *Vital Speeches of the Day*, November 1, 1989. Argues that the United States and Canada must cooperate in reducing acid rain.

Edward Edelson, *Clean Air*. New York: Chelsea House, 1992. Discussion of causes and solutions of air pollution problems, including acid rain.

Environmental Protection Agency, *Acid Rain: A Student's First Sourcebook*. Washington, DC: Environmental Protection Agency, 1990. Projects and information for studying and remedying acid rain.

Ned Helme and Chris Neme, "Acid Rain: The Problem," *EPA Journal*, January/February 1991. Argues that acid rain is still a serious problem and must not be ignored.

J. Laurence Kulp, "Acid Rain: An Environmental Problem?" *The World & I*, November 1991. Gives evidence from studies that show acid rain is not a serious problem and argues that acid rain can be remedied with significantly less cost than that proposed by the 1990 Clean Air Act.

Edward Krug and W.L. Warnick, "Sources of Acidity in Surface Waters." *Science* 253 (1991). Study showing lake acidification is caused by the shape of surrounding land and the type of surrounding vegetation, not by acid rain.

Eileen Lucas, *Acid Rain*. Chicago: Childrens Press, 1991. Discusses what acid rain is, how it damages the environment, what causes it, and ways to help prevent it.

Reed McManus, "Ambushed from Within," *Sierra*, September/October 1992. Contends that President Bush tried to sabotage the Clean Air Act.

Dixy Lee Ray with Lou Guzzo, *Trashing the Planet*. Washington, DC: Regnery Gateway, 1990. Discusses all the major environmental "crises," including acid rain, and argues that they are not crises at all.

Cheryl Simon Silver with Ruth S. DeFries, *One Earth, One Future: Our Changing Global Environment*. Washington, DC: National Academy Press, 1990. Describes the various environmental problems facing the planet, including acid rain, which they contend is a serious problem.

USA Today, "Acid Rain: Forests May Not Be in Danger," June 1991. Evidence shows that acid rain is not damaging commercial forests.

Utne Reader, "Acid Rain—One of America's Worst Killers," September/October 1989. Quotations and statistics that point to acid rain as an indirect cause of many cancer cases.

ORGANIZATIONS TO CONTACT

The following organizations are concerned with the issues debated in this book. All of them have publications or information available for interested readers.

The Acid Rain Foundation
1410 Varsity Dr.
Raleigh, NC 27606
(919) 828-9443

The foundation was established to raise the level of public awareness about acid rain. It publishes the quarterly newsletter *Update*, the yearly *Resource Directory*, and all sorts of educational materials for schools and students.

Air Resources Information Clearinghouse
99 Court St.
Rochester, NY 14604
(716) 546-3796

Formerly called the Acid Rain Information Clearinghouse, this organization gathers, organizes, and distributes information on acid rain and other issues concerning air quality. It publishes the *Acid Rain Digest* monthly.

Alliance for Acid Rain Control
444 N. Capitol St., Suite 256
Washington, DC 20001
(202) 624-5475

The alliance is a group of U.S. governors concerned about acid rain. It lobbies Congress to pass acid rain laws that give benefits worth the amount of money to be spent.

Electric Power Research Institute (EPRI)
3412 Hillview Ave.
Palo Alto, CA 94303
(415) 855-2000

EPRI conducts research on how to produce electricity in ways that are not too costly for the economy or the environment. It is a good source for statistics and studies on acid rain. It publishes its *Journal* nine times a year and the *Guide* three times a year.

The Heritage Foundation
214 Massachusetts Ave. NE
Washington, DC 20002
(202) 546-4400

This conservative research organization has many scholars working on various issues and policies important to the United States, including environmental issues such as acid rain. Its position on issues tends to favor limited spending and limited government interference in public affairs. The foundation opposed the acid rain amendment to the Clean Air Act. It publishes a monthly journal titled *Policy Review* as well as many papers and lectures throughout the year.

National Coal Association
1130 17th St. NW
Washington, DC 20036
(202) 463-2625

The association represents the coal industry's interests. It lobbies Congress to support laws that favor the coal industry. It believes the acid rain law is too expensive and will hurt the coal industry and the U.S. economy. It publishes the weekly *Coal News* and the bimonthly magazine *Coal Voice*.

INDEX